DINOSAURS

VELOCIRAPTOR

BY GOLRIZ GOLKAR

Kids Core

An Imprint of Abdo Publishing
abdobooks.com

abdobooks.com

Published by Abdo Publishing, a division of ABDO, PO Box 398166, Minneapolis, Minnesota 55439.

Printed in the United States of America, North Mankato, Minnesota.
052025
092025

THIS BOOK CONTAINS RECYCLED MATERIALS

Cover Photos: Sebastian Kaulitzki/Shutterstock Images (Velociraptor); Shutterstock Images (background)
Interior Photos: Daniel Eskridge/Stocktrek Images/Getty Images, 4–5; Shutterstock Images, 6, 16; Lucas Vallecillos/VW PICS/Science Source, 8; Sergey Krasovskiy/Stocktrek Images/Science Source, 9; Daniel Eskridge/Shutterstock Images, 10; Mohamad Haghani/Stocktrek Images/Science Source, 12–13; Richard Jones/Science Source, 15; iStockphoto, 18; Danny Ye/Shutterstock Images, 20–21; Mark Stevenson/Stocktrek Images/Science Source, 23; Red Line Editorial, 24; Universal Pictures/Photofest, 26; Sebastian Kaulitzki/Science Photo Library/Getty Images, 28–29

Editor: Kari Cornell
Series Design: Mary Shaw

Library of Congress Control Number: 2024948982

Publisher's Cataloging-in-Publication Data

Names: Golkar, Golriz, author.
Title: Velociraptor / by Golriz Golkar
Description: Minneapolis, Minnesota: Abdo Publishing, 2026 | Series: Dinosaurs | Includes online resources and index.
Identifiers: ISBN 9781098297367 (lib. bdg.) | ISBN 9798384919889 (ebook)
Subjects: LCSH: Velociraptor--Juvenile literature. | Dinosaurs--Juvenile literature. | Carnivorous animals--Juvenile literature. | Paleontology --Juvenile literature. | Extinct animals--Juvenile literature.
Classification: DDC 568.19--dc23

CONTENTS

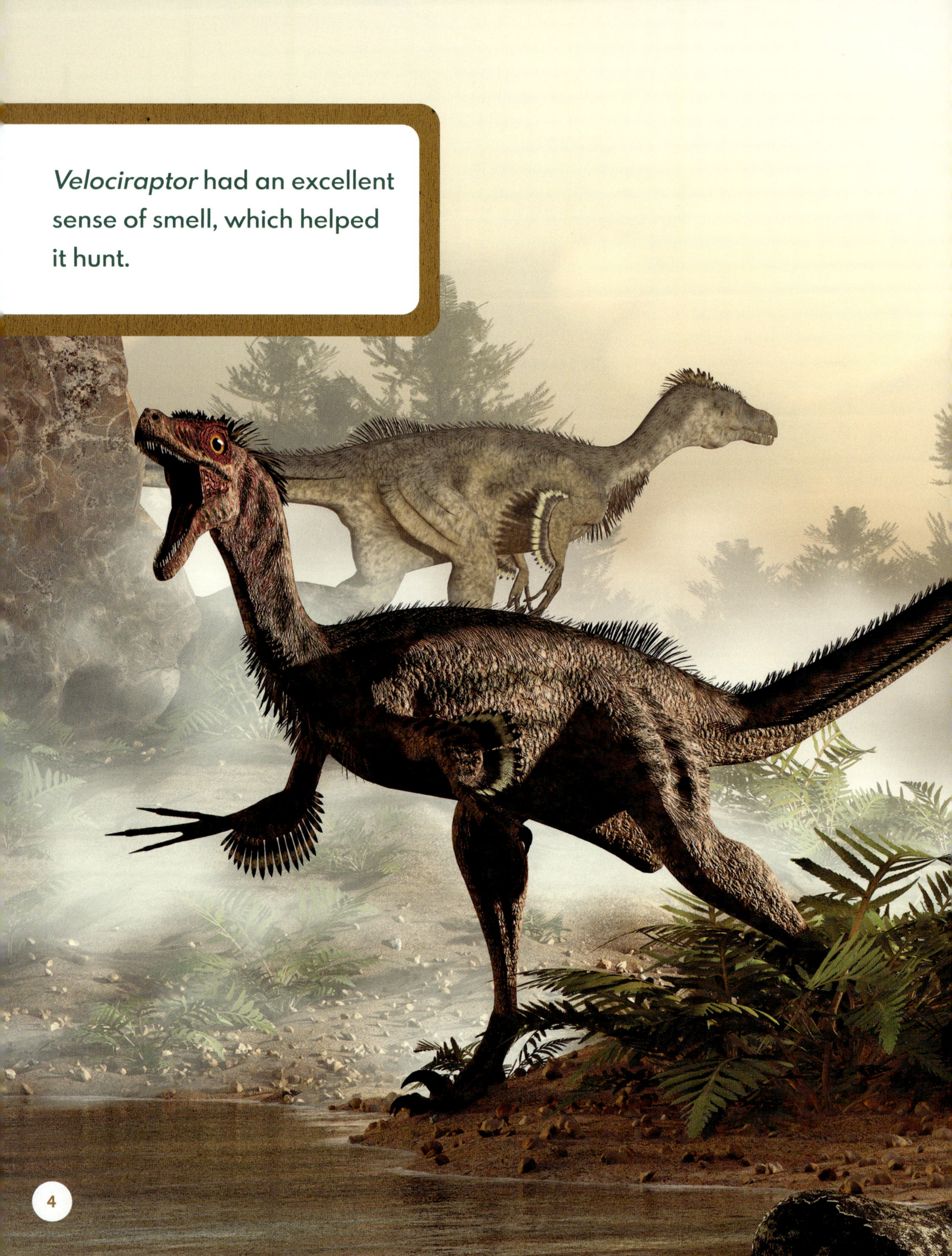

Velociraptor had an excellent sense of smell, which helped it hunt.

CHAPTER 1

SMALL BUT SWIFT

A *Velociraptor* (veh-LAH-sih-rap-ter) sniffs around in the desert. It is looking for its next meal. Suddenly, a breeze blows in its direction. The *Velociraptor* turns its head and picks up a scent. It spots a young *Protoceratops*.

Scientists believe *Velociraptor* had feathers on its arms.

The *Protoceratops* is using its beak to rip leaves off bushes.

The *Velociraptor* races to the young dinosaur on its two big hind legs. It grabs the **prey** with its winglike arms. It digs into the *Protoceratops* with its **sickle**-shaped claw. The *Protoceratops* tries to bite the *Velociraptor* with its beak. But it's too late. The *Velociraptor* sinks its jagged teeth into its prey.

When Dinosaurs Roamed

Dinosaurs lived during the Mesozoic Era. This was 252 million to 66 million years ago.

Velociraptor Brains

Velociraptor had a fairly large brain. The brain helped it keep its balance. It also allowed these dinosaurs to make difficult movements when grabbing prey. They could attack another dinosaur by jumping on it with both feet. They could also grab a dinosaur with their front claws while stabbing it with their hind legs.

A life-size model of *Velociraptor*'s footprint is displayed at the Karpin Fauna wildlife center in Spain.

Earth had one large landmass back then. Earthquakes and volcanic eruptions eventually divided the land into several smaller **continents**. Changes in climate and available food led to the **evolution** of different dinosaurs.

Scientists believe *Velociraptor* was smart due to its large brain.

The Cretaceous period was at the end of the Mesozoic Era. This was about 145 million years ago. Dinosaurs called dromaeosaurs appeared.

Velociraptor had four claws on each foot. One claw was small and located away from the other three. Another claw was very sharp and used for hunting.

They lived in the deserts of central and eastern Asia and in North America. These large-headed dinosaurs walked upright on two feet. They had specialized claws and wrist joints. These helped them grab and kill prey. *Velociraptor* evolved from this family of dinosaurs.

Explore Online

Visit the website below. Does it give any new information that wasn't in Chapter One?

The Dromaeosauridae

abdocorelibrary.com/velociraptor

Velociraptor's powerful back legs allowed it to run quickly.

A MIGHTY HUNTER

Velociraptor was up to 6 feet (1.8 m) long. It weighed up to 100 pounds (45 kg). It was about the size of a turkey. These dinosaurs looked like unusual birds. They had two winglike arms and two clawed feet.

Their bodies were covered in feathers. But they could not fly. Their arms were not long enough. The feathers might have helped them attract mates. Or the feathers may have helped them keep nest eggs warm.

Velociraptor had a long tail with strong muscles. It helped the dinosaur balance while attacking prey. The tail also helped the dinosaur steer in the right direction while running. *Velociraptor* could run faster than a human. It could go faster than 25 miles per hour (40 km/h)!

A Skilled Hunter

Velociraptor was a skilled **predator**. Its large brain helped it plan and carry out attacks

Dromaeosaurs, the family of dinosaurs from which *Velociraptor* evolved, likely laid eggs in nests. The dinosaurs' feathers may have helped keep the eggs warm.

on other animals. The dinosaur also had large nostrils. This made it easy to smell prey. *Velociraptor*'s teeth faced backward. This design helped it hold on to prey.

Velociraptor had about fifteen sawlike teeth in its upper jaw and the same number in its lower jaw.

Velociraptor had a sickle-shaped claw on each foot. Each foot had four claws. Three claws remained on the ground as it walked. The sickle claw stayed up in the air. This helped keep it sharp. *Velociraptor* used this claw as a weapon. It stabbed and hooked its prey before eating it.

Velociraptor fossils are usually not found together. This means that these **carnivores** most likely hunted on their own. They ate small animals, including lizards and mammals. They ate young dinosaurs and dinosaur eggs.

Dinosaurs and Modern-Day Birds

Today's birds evolved from two-legged dinosaurs such as *Velociraptor*. *Velociraptor* had many birdlike features. Its wrists moved in a flapping motion. It folded its arms against its body like wings. *Velociraptor* had feathers and forward-facing toes. The sickle claw was similar to the long claws of hawks and eagles.

Velociraptor used its razor-sharp claws to dig into prey such as small reptiles, amphibians, and smaller dinosaurs.

Some *Velociraptor* fossils suggest it was also a **scavenger**. Bits of animal remains were found in their stomachs. These findings helped scientists choose their name. *Velociraptor* means "quick thief" in Greek. This dinosaur likely stole other animals' meals.

Further Evidence

Look at the website below. Does it give new evidence to support Chapter Two?

National Geographic Kids: Velociraptor

abdocorelibrary.com/velociraptor

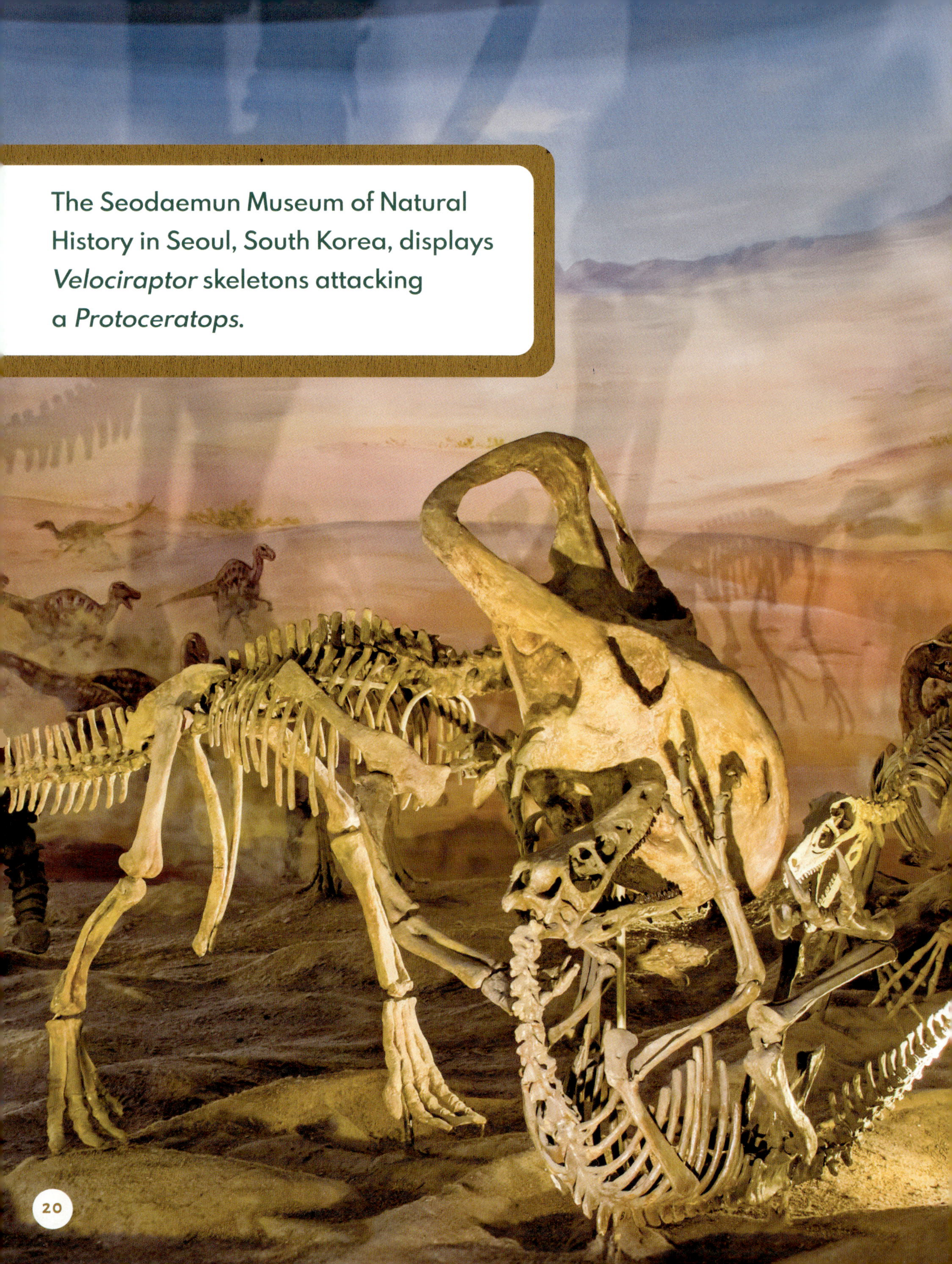

The Seodaemun Museum of Natural History in Seoul, South Korea, displays *Velociraptor* skeletons attacking a *Protoceratops*.

CHAPTER 3

DISCOVERING VELOCIRAPTOR

The first *Velociraptor* fossils were found in 1923 in Mongolia. They included a claw and a skull. Many years later in 1971, scientists made another discovery. They found *Velociraptor* remains tangled up with the fossil of a *Protoceratops.*

Paleontologists think the two dinosaurs were fighting when they died. The *Velociraptor*'s claw was in the other dinosaur's throat. The mouth of the *Protoceratops* grasped the *Velociraptor*'s arm. A mound of sand fell and buried the dinosaurs.

The Fighting Dinosaurs

Scientists believe that dinosaurs rarely attacked much larger dinosaurs. It's possible the *Velociraptor* attacking the *Protoceratops* was very hungry. It may have been young and unskilled at hunting as well. Other *Protoceratops* fossils have shown raptor bite marks. *Velociraptor* probably fed on *Protoceratops* carcasses. It's unlikely it attacked the dinosaurs.

Protoceratops fossils have been found with slashes that appear to have been made by *Velociraptor* teeth.

Another fossil shows a bone from a pterosaur in a *Velociraptor*'s stomach. The pterosaur was a much bigger reptile. This means *Velociraptor* likely scavenged the bone from its remains.

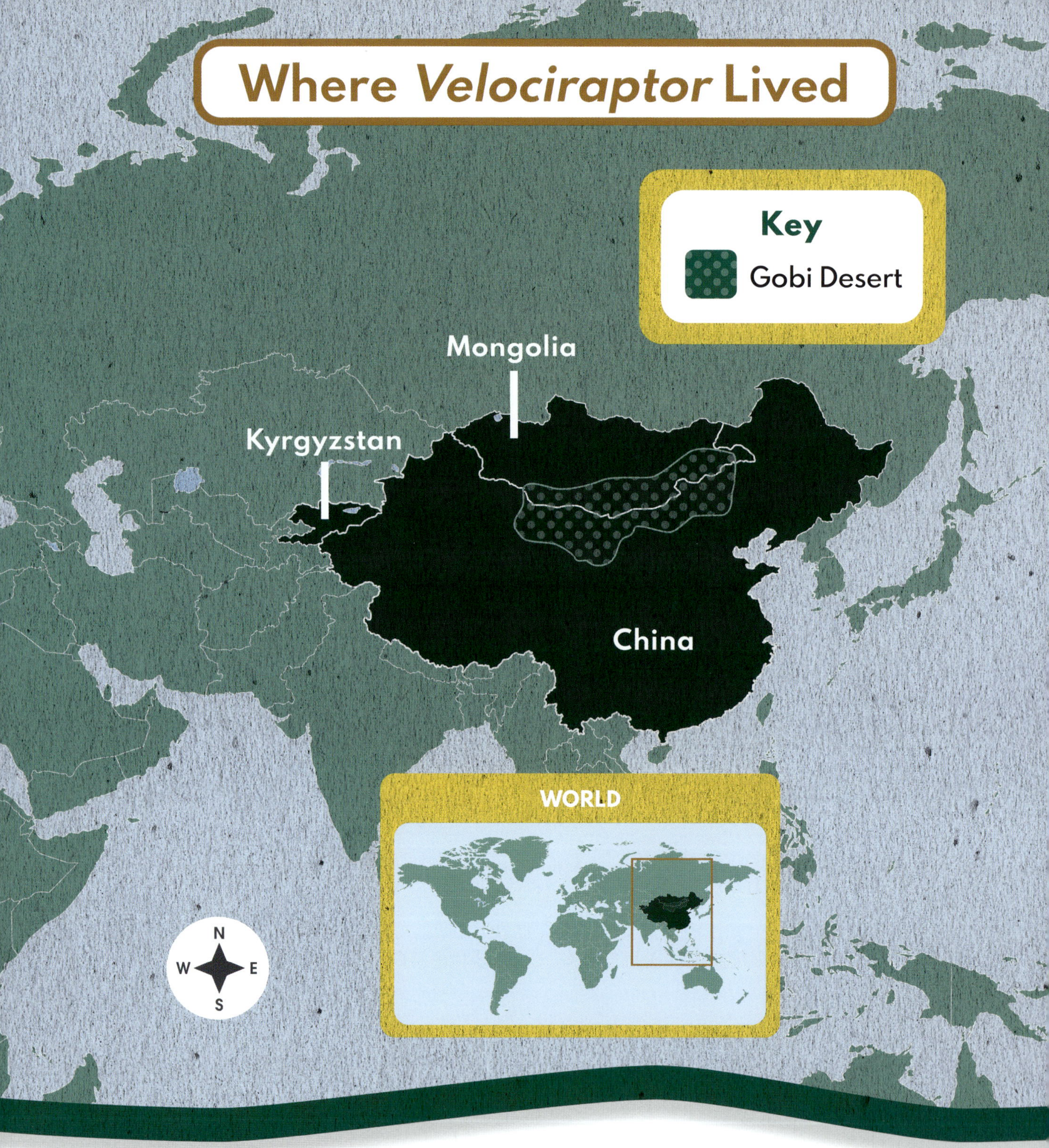

This map shows where *Velociraptor* once lived. Fossils have been found in the Gobi Desert in China and Mongolia and in Kyrgyzstan.

Another fossil discovery in China in 1999 led paleontologists to a different type of *Velociraptor*. They noticed this one had a different kind of jaw. Many *Velociraptor* fossils are displayed at the American Museum of Natural History in New York. Others are at the Australian Museum.

In Popular Culture

Velociraptor has appeared in the popular *Jurassic Park* movies. It is shown as a fierce predator. *Velociraptor* is also shown hunting in packs. Scientists don't believe this was true.

Velociraptor also appeared in *The Good Dinosaur* and *Dinosaur Train*. Video games such as *Lego Jurassic World* show *Velociraptor* too.

Velociraptor appeared in the movie *Jurassic World*, released in 2015. The dinosaurs in the movie were missing their feathers. They also were larger than in real life.

By studying *Velociraptor* fossils, scientists have learned that it was a small but swift dinosaur. It was always on the hunt for its next meal.

PRIMARY SOURCE

Dinosaur researcher Dr. David Button discusses *Velociraptor*:

> I think . . . of *Velociraptor* like a land eagle as they're very similar to eagles in many ways and . . . they would have behaved similarly as well.

Source: Lisa Hendry. "Vicious *Velociraptor*: Tales of a Turkey-Sized Dinosaur." *Natural History Museum*, n.d. nhm.ac.uk. Accessed 18 Sept. 2024.

Comparing Texts

Think about the quote. Does it support the information in Chapter Three? Or does it give new information? Explain how in a few sentences.

DINO DETAILS

Long tail to help move and keep balance

Two large hind legs for running

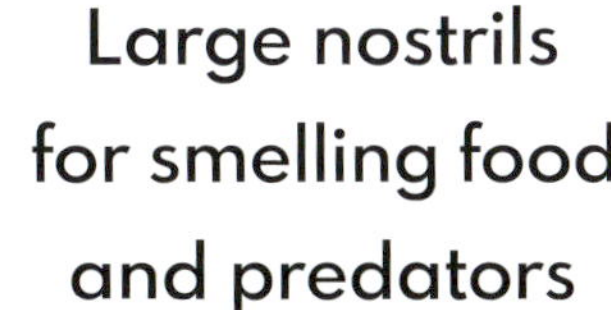
Large nostrils for smelling food and predators
Jagged teeth for holding on to prey
Swivel-jointed wrists for snatching prey
Claws on each foot, including one sickle claw per foot for grabbing prey

Glossary

carnivores
animals that eat meat

continents
the world's seven major landmasses

evolution
a change in species over time

paleontologists
scientists who study fossils

predator
an animal that hunts other animals

prey
an animal who is food for a predator

scavenger
an animal that feeds on dead animals

sickle
a curved object used for cutting

Online Resources

To learn more about *Velociraptor* and late-Cretaceous dinosaurs, visit our free resource websites below.

Visit **abdocorelibrary.com** or scan this QR code for free Common Core resources for teachers and students, including vetted activities, multimedia, and booklinks, for deeper subject comprehension.

Visit **abdobooklinks.com** or scan this QR code for free additional online weblinks for further learning. These links are routinely monitored and updated to provide the most current information available.

Learn More

Carr, Aaron, and John Willis. *Velociraptor.* Weigl, 2021.

Lim, Angela. *Archaeopteryx.* Abdo, 2026.

Woodward, John. *The Dinosaur Book.* DK, 2023.

Index

About the Author

Golriz Golkar has written more than 100 nonfiction and fiction books for children. She holds a B.A. in American literature and culture from UCLA and an Ed.M. in language and literacy from the Harvard Graduate School of Education. Golriz lives in France with her husband and young daughter.